AF785067

ORCHIDS

BY KIM BOGREN OWEN
PHOTOGRAPHY BY AMBER OWEN, RON HANKO, AND MIKE PEARSON

Dedicated to my grandfather, Dale Baker, for passing on a love of growing things

Copyright © 2016 Kim Bogren Owen
Photographs copyright © 2016 Amber Owen, Ron Hanko, Mike Pearson

Library of Congress Number
2016900330

ISBN
9780997200713

First Printing 2016
Printed in the United States

No part of this publication may be reproduced in whole or in part, or stored in a retrieval system, or transmitted in any form or by any means, electronic, mechanical, photocopying, recording, or otherwise, without written permission of the publisher. For information regarding permission, email wordsreflected@gmail.com.

All photographs by Amber Owen, except pages 15 and 20 by Ron Hanko, and page 21 by Mike Pearson

Like people, orchids come in all shapes and sizes.

Some have
leaves that reach
for the sun,

others have
leaves that hug
the ground.

Some have
flowers that grow
on a single stem,

others grow in clusters of flowers.

Some have large flowers,

others have
small flowers.

Some grow in
cool places,

others grow in hot places.

Orchid flowers come in many colors — pink, red, black,

yellow, white, and even green!

Some even have striped flowers!

But like other flowers, all orchids attract birds, bats, and insects to pollinate them.

Pollination helps flowers make seeds, so new plants can grow. But orchids use special tricks to attract just the right pollinator.

Orchid flowers are just the right size and bloom at just the right time for their special pollinator to be there.

This *Lepanthes* looks just like a female gnat to trick the male into pollinating it.

People pollinate
one orchid—
Vanilla! That is
so we can use
the seedpods in
oatmeal, cookies,
ice cream, and
many other foods.

Even though orchids come in all shapes and sizes, what makes them really special is that all their flowers are symmetrical.

That means they are the same on both sides. That is a lot like people's faces, even yours!

Parents, here are some activities you can do with your child as you read this book to expand on the concepts, vocabulary, and pictures.

* Find things in your house that match the vocabulary and expand on it. For example, find things that are large and small, different colors, or symmetrical.

* Talk about the plants in your house, yard, or park that grow in clusters, or on a single stem. You can also talk about fruits that grow in clusters, like grapes or bananas.

* Spend some time looking at the photographs and ask your child to tell you about what they are looking at.

* Count the number of petals or leaves in a picture and find something in your house with that same number.

* Have your child curl up in a small ball like a closed flower and then slowly open up to "bloom."

* Make a symmetrical painting by painting on one side of a piece of paper and then folding it in half. When you open it up, ask your child to describe how it is symmetrical.